WHITCHURCH AIRPORT

WHITCHURCH AIRPORT

An Account of the Early Days of Flying in the Bristol Area

Gerald S. Hart

Crockerne Books
Pill North Somerset
1997

First published 1997 by
Crockerne Books 95 Westward Drive Pill Bristol BS20 0JS

ISBN 0 95160741 1 3

Also by author
"Ministry 200 - The story of Pill Union Church"
"Ham Green"

Typeset with Amstrad PC7486 SLC 33 using Microsoft Works for Windows
and Epsom Stylus Printer 800

Printed and Bound by Antony Rowe Limited Chippenham

Contents

ILLUSTRATIONS

Illustrations (cont.)

The originals or negatives of the pictures on pages 6, 11, and 12 were not available to the author. Photo-copies have been used. The reproductions are certainly not up to the standard expected in a modern production. However, they are nearly seventy years old and have been included for historical reasons.

AUTHOR'S INTRODUCTION AND ACKNOWLEDGEMENT

Most readers would expect a book on aviation to be written by someone who has had considerable experience in that field. It is not the case with this particular booklet. True it contains a sense of nostalgia from the author's memories of a six-month spell as a traffic assistant at Whitchurch in 1936; he was eighteen at that time, but there his connection with matters aeronautical ended. Even flying as a passenger in those early days was limited to one trip in a Western Airways Dragon from Whitchurch to Weston-super-Mare. There followed a period of fifty years when he did no flying, perhaps, remedied by fairly extensive holiday flying during the last ten years. That has included six crossings of the Atlantic by air, certainly not exceptional in these days. The main reason for this work is a love of local history. It is based on a project carried out in the first year of a study on local and regional history at the University Sector College, Bath College of Higher Education. However, the project has been revised to make the reading more "anecdotal", but, all references and details of sources have been retained.

In acknowledging all the help that is so readily given to many local history researchers it cannot be forgotten that the work can result in the making of many new friends and associates, even when the link has only been on the phone. Such has been the case in the study for this publication. The writer thanks the tutorial and library staff of the Bath College of Higher Education, the staff of the former Avon libraries at Filton, Bristol Central, Portishead and Pill, the staff of the Bristol Record Office, Gary Brook of the Evening Post Library, and Mike Fish, archivist British Aerospace. Such professional staff have the skill to recognise those who are seriously interested in a subject and go out of their way to help.

To others a, "thank you" has to be said for almost sharing the work. Mention must be made of the swapping of information with Shelley Peterson of Hartcliffe School, she has made it possible for her students to study Whitchurch Airport as a local history subject for their GCSE examinations. The help of local transport historian, Michael Tozer, is readily acknowledged. Conveniently living a few yards away from the author, he has produced material from his vast collection of local pictures, books and cuttings from magazines and newspapers. All the pictures of restored aircraft have been given me by Edwin Shackleton. He is a retired aerospace test

Author's Introduction and Acknowledgement (cont.)

engineer with the hobby of flying as a passenger in as many types of aircraft as he can. His 1997 entry in the Guinness Book of Records is the eighth and records a total of 600 different planes and other aircraft. For an author-produced book to be created much help is required. Thanks go to Maureen de Lancey for the tedious task of reading the script, and to Richard Byles, who has acted as the writer's mentor for many years in matters photographic. Thanks also to fellow-student, Ken Andrews for his ready advice when that modern beast, the computer, has not been fully co-operative.

It must be said that there is gratitude for the work of those who wrote the many books and papers in the list of references and bibliography, reading them was pleasurable. Finally permission to use material is gratefully acknowledged from Edwin Shackleton, the Bristol Record Office, the Evening Post, Michael Tozer and Aer Lingus.

BRISTOL AIRPORT

Details of this air photograph, taken before the Second World War, can be seen on the plan on page 22. The concrete circle in the top left hand corner was used for making adjustments to the compass. The plane would have been orientated on the circle, magnetic variation for the date etc. would have been known, the "deviation" of the compass, due to ferrous metal in the plane itself, would then be estimated at various compass positions. In those days that would have been mainly caused by the engine. Photograph M. J. Tozer collection.

To those courageous pioneers of flying in the Bristol Area.

BRISTOL ENTERS THE FIELD OF AVIATION

In 1903 brothers Orville and Wilbur Wright made the first ever flight in a machine that was heavier than air; as early as 1909 Louis Bleriot flew across the English Channel; and in just one year, 1910, Bristol became involved in aviation. The man mainly responsible was Sir. George White. He was already connected with transport, Bristol's trams and buses, and used redundant tram sheds at Filton, just beyond the northern boundary of the city, to start a factory for the manufacture of aeroplanes. His company was called, "The British and Colonial Aeroplane Company Limited. The first aeroplane, the Bristol Boxkite, was produced within months of starting and by the end of the first year an order for eight had been received from Russia.[1] The design and construction of aircraft continued in a very successful way, in the military field the Bristol Scout became an important reconnaissance plane and the Bristol Fighter was claimed to be, "among the most famous military aeroplanes ever produced".[2] About 5,300 were built for the First World War and by the end of that conflict the 200 workers of 1914 had risen to 3,000. The company acquired the Cosmos Engineering Company in 1920, enabling it to produce aero-engines, and in the same year it changed its name to, "The Bristol Aeroplane Company".

The inclusion of the word, "Colonial" in the first title of the company was very apt, in fact the activities were not just confined to the colonies but were also world-wide. In those early days it was also realised that if planes were to be sold pilots would have to be trained. The Company had two flying schools, one at Larkhill and the other, with other firms, at Brooklands. The government had a programme whereby volunteers from the services who paid for their own training were given £75 when they passed. The Bristol Company trained more than half the pilots who qualified under that scheme.[3]

Sir George White died in 1916; he lived long enough to see the start of his enterprise make a major contribution to the defence of his country, maybe he foresaw its ultimate success.

One of the great aircraft designers of that day was Frank Barnwell, he had left Filton at the beginning of the war to serve in the Royal Flying Corps, but was "persuaded" to return and set up a secret drawing office in No 4 Fairlawn Avenue, one of the delightful Edwardian workers' houses built within the grounds of the Filton complex.[4] Barnwell had already been mainly responsible for the design of the single seater, Bristol Scout, and here as Chief Designer the Bristol Fighter had its birth.

The early airfield at Filton grew to an Aerodrome, necessary for so large an aircraft factory; the only other airfields near Bristol were at Yate and on the Duke of Beaufort's estate at Badminton. Filton Aerodrome became the home of the RAF Reserve No 501 (County of Gloucester) Squadron in June 1929. By 1930, because of local connections, it changed its name to the No 501 (City of Bristol) Squadron. Their first Commanding Officer was Flight Lt. L. P. Winters, a regular RAF officer.[5]

Between the wars, during the recession, work was not easy at Filton, however, there was a demand for planes in the civil and owner-flier fields and the new Bristol Bulldog was a successful military machine. Lord Rothermere, the newspaper magnate, offered to sponsor what he wanted to be the world's fastest air-liner, with the condition that it had to be built in one year.

Filton had on the drawing board an eight passenger version of its twin Aquila135 project, popularly called after Frank Barnwell, using his military rank, "The Captain's Gig". The offer was half the cost of £18,500 to start and the balance if completed in a year. The first cheque was paid immediately and the work proceeded. Known as the Bristol 142, it was called, "Britain First" by Lord Rothermere who presented it to the nation. It never became a commercial passenger plane and was not allowed to leave the country. It was of course the forerunner of the Bristol Blenheim Bomber.[6] Chief designer Frank Barnwell died in an air crash in 1938.

John Pudney, in the last pages of his book, "Bristol Fashion", hints at regret that the early work at Filton was so much involved in war. He quotes Horace Walpole writing towards the end of the eighteenth century that he did not care to spend his old age, "divining with what airy vehicles the atmosphere will be peopled hereafter, or how much more expeditiously the east, west or south will be ravaged and butchered". He also draws attention to Winston Churchill's prophesy of, "the crash of bombs exploding in London and the cataracts of masonry and fire and smoke". He also suggests that Winston Churchill, "expressed the regrets of civilised man", when in the House of Commons, in February 1934, he referred to, "this cursed, hellish invention and development of war from the air has revolutionised our position. We are not the same kind of country we used to be when we were an island, only twenty years ago." This project is intended to be about aviation as commercial transport, alas, it is difficult to separate it from what many would see as the misuse of flying, albeit much of that is as a defence. However in our own time Bristol and Filton have made major contributions to commercial flying, both in the aircraft and the engines that power them, culminating in "Concorde", the pride of the people of Bristol and its surroundings, and with other European countries the development and production of the series of large passenger planes under the collective title of "Air Bus".

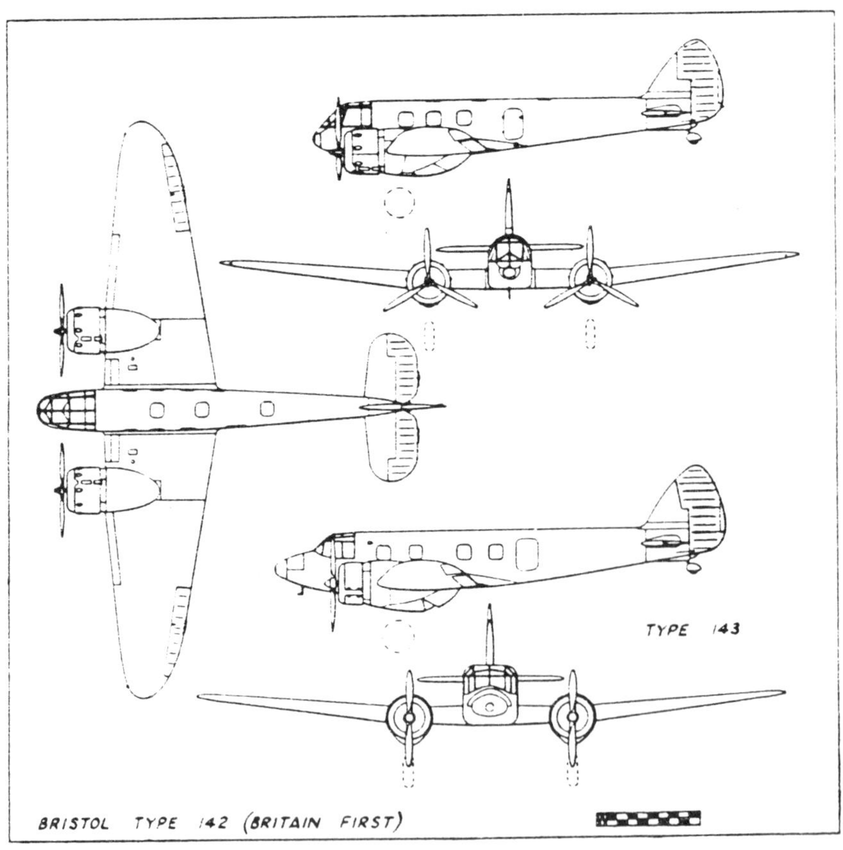

Some early plans of the passenger plane that became the Bristol Blenheim Bomber.

WHITCHURCH IN SIGHT

As early as 1910 Bristol had an Aero Club, The Bristol and West of England Aero Club. However, it did not continue after the outbreak of the 1914 to 1918 War. Most of its activities were confined to gliding and model making. One member, a schoolboy Norman Edgar, who was to become very important in local commercial aviation, designed and built a glider with a motor-cycle engine. It did not work, fortunately there were no accidents.[7]

Towards the end of the 1920's a common method of learning to fly was to join a local aeroplane club. Such a club was considered for the Bristol area and a public meeting was called for February 1927; their support was promised by several leading citizens. Help was not forthcoming from the Air Ministry, in fact its advice to the meeting was that the grant of a subsidy was not possible, and even worse, no club could run without one.[8] Nevertheless, it was decided to go ahead as George Parnall offered not only the use of his airfield at Yate but also a Parnall "Pixie" light aeroplane and a cash donation. The first committee was chaired by Mr. A. H. Downes-Shaw, thus the Bristol and Wessex Aeroplane Club was born. It is still in existence at the Bristol Airport at Lulsgate, having managed to overcome recent financial problems. The Bristol Aeroplane Co., Ltd. gave help in arranging an early flying meeting at Filton and there was extensive fund-raising. The first Pilot Instructor,

Mr. E. B. W. Bartlett, who had served during the 1914 - 18 War, was appointed. Filton Aerodrome was seen as, "one of the most accessible in the whole of the country",[9] the Air Ministry was influenced by what was going on at Bristol and changed its mind. The use of Filton Aerodrome for a year was offered to the new Club, its recognition by the government also enabled the Club to claim grants, up to £2,000 per year, for all pilots trained. Encouraged by George Parnall, in spite of his earlier generous offer, the Club accepted.[10]

MISS W. S. BROWN, A COMPETITOR AT THE WESSEX AIR PAGEANT.

Many ladies took part in those early days of flying.

There followed great activity by the Aeroplane Club. In 1928 there were four "official" air pageants, the Bristol and Wessex being chosen to organise the Western Area Pageant. Some

30,000 people attended the event at Filton on 5th May, no doubt the vast majority of those attending used George White's trams to get to Filton. Photographs show the numbers of aircraft that were present, local dignitaries that attended and many, men and women, famous people in aviation at that time who were present. Certainly the Air Pageant placed Bristol firmly on the civil aviation map and showed the members of the Bristol and Wessex Club as leaders for the expansion of aviation in the West Country. A year from its start the Club was exceedingly prosperous, it had as many as seventy pilot members, a flying school, social premises at Filton and a large "social" membership. It possessed four modern aeroplanes, with the aim to "bring aviation within reach of everyone, including those with only moderate means". It planned to expand its membership to people living in Bath, Gloucester, Cheltenham, Taunton and Bridgwater, and "With an increased turnover, it [was] hoped to be able to reduce charges for solo practice very considerably".[11] The Club was very happy at Filton, it had good relations with the Bristol Aeroplane Co., Ltd. and its officers and pilots; all appeared to be well.

The first concern of the Club was the establishment by the Government of the organisation, "National Flying Services Limited". It was intended to make one central grant to enable the new group to arrange pilot training throughout the country and thus save money on grants. The scheme floundered.[12] Nevertheless, there arose the problem of the presence at Filton of the new RAF Squadron. True the Club had the time for the use of Filton extended to three years, but it was known that the use of the same aerodrome by the RAF and a local flying club could lead to problems. The committee of the Club, with the aid and approval of the Air Ministry, looked elsewhere for a new home. A site at Whitchurch, with the advantage of being nearer the centre of Bristol than Filton, was approved by the Air

Ministry. The disadvantage of the near proximity of Dundry Hill and its high church tower was taken into account; nevertheless, the Air Ministry considered Whitchurch to be the best position for a Bristol Airport. The Aero Club obtained options on the purchase of the land involved, some 290 acres. The proposed site would allow a landing and take-off length of 1,000 yds in any direction; that was one of the requirements of the Air Ministry at that time.

This was a time when municipal airports were being considered, Section 8 of the Air Navigation Act allowed local authorities to build them. The Government of the day encouraged it. During those early years of aviation in the late 20's the Central Government was Conservative with Stanley Baldwin as Prime Minister. After the election of 1929 the Labour Party was the party with the most seats. Ramsay Macdonald became Prime Minister but his party was in office but not in power. The National coalition was formed in 1931 and confirmed by a General Election. Ramsay Macdonald continued as Prime Minister, he was the first British Prime Minister to use an aeroplane for official duties.[13] In the local situation the Bristol Council was in the control of the "Citizens' Party", a "ratepayers" party, certainly the Labour opposition considered they were facing a Tory - Liberal coalition.[14] There is little doubt that the central and local governments, in those days of great depression, very much influenced the advance of civil aviation, particularly as grants to relieve unemployment were available. Nevertheless, full credit must go to the Bristol and Wessex Aeroplane Club for its initiative in seeing the importance of a civil airport. However, the Club did have the advantage of the creation of a home for their social and flying activities without a large capital outlay. It must also be stated that the Bristol Council was very much in favour of being in the van of the provision of facilities throughout the country for civil aviation. As early as 13th September 1927, a councillor, Mr. Lyne, asked a question in the council chamber regarding what Bristol was doing to afford facilities for flying, particularly for owners of private planes. He stressed the importance of the matter if Bristol was to be in line with modern trends. He must have been delighted with the reply from the Chairman of the Planning Committee to the effect that not only had Bristol consulted with neighbouring authorities, but a reply from the Air Ministry was awaited for a meeting with officials to consider sites.[15] Mr. Lyne was to become one of the first members of the Bristol Airport Committee.

The great decision was made by the Bristol Council at its meeting on 14th May 1929, when it decided to accept, and act on, a report on a municipal airport from the Planning Committee. It was agreed to accept the offer from the Aeroplane Club of the option to purchase land at Whitchurch for the sum of £16,386. Because of a large cost saving it was also agreed for the Bristol and Wessex Aeroplane Club to manage the airport on behalf of the Council. It was also realised that it would be necessary to extend the boundaries of the City and County of Bristol to include the aerodrome site.[16] As well as the saving by allowing the Aeroplane Club to manage the airport, often called the aerodrome in the minutes, the Council was to receive £100 a year from the Club for the use of the purpose built clubhouse. It was also decided to form a Council Airport Committee, the first one consisting of Aldermen Curtis, Dowling, Sennington, and Woodcock, with Councillors Berriman, Griffiths, Jones, Lyne, A.L.W. Smith and Wroe. There were also plans for a new road as the only access to the airport site was the lane from Whitchurch to Bishopsworth.

The massive task of clearing the site and building the clubhouse, showroom, and hangar with the necessary approach drives and aprons proceeded. On the 9th July 1929, following a report to the Council, increased expenditure was approved. To create work for the unemployed, hedges were grubbed out by hand instead of machinery, unfortunately an expected grant from the government towards this work was halved.[17] The plan was for the aerodrome to be ready by January 1930 for the Club to move in. Appalling weather prevented this so an extension of time at Filton for the Aeroplane Club had to be obtained from the Air Ministry. At that time when work was not possible because of rain, or other bad weather, workers had to stand by without payment. At one stage in December, to enable the labourers to obtain unemployment benefit, the work was stopped for a week.[18] There were other problems including the objection at a town meeting to the closure of footpaths that crossed the site. When the Club did make its move to Whitchurch members were full of praise for the improved facilities.

SIR SEFTON BRANCKER EXPLAINING HIS MOTH TO THE LORD MAYOR OF BRISTOL (ALDERMAN AND MRS. CURLE.)

[*Bristol Times and Mirror*

BRISTOL AND WESSEX AERO CLUB: THE DUKE OF BEAUFORT, MR. TALBOT O'FARRELL, THE DUCHESS OF BEAUFORT, COL. H. C. WOODCOCK, M.P., CAPT. A. S. C. BROWNE AND MR. A. H. DOWNES-SHAW.

Bristol and Wessex Aeroplane Club Pageant at Filton, 5th May 1928.

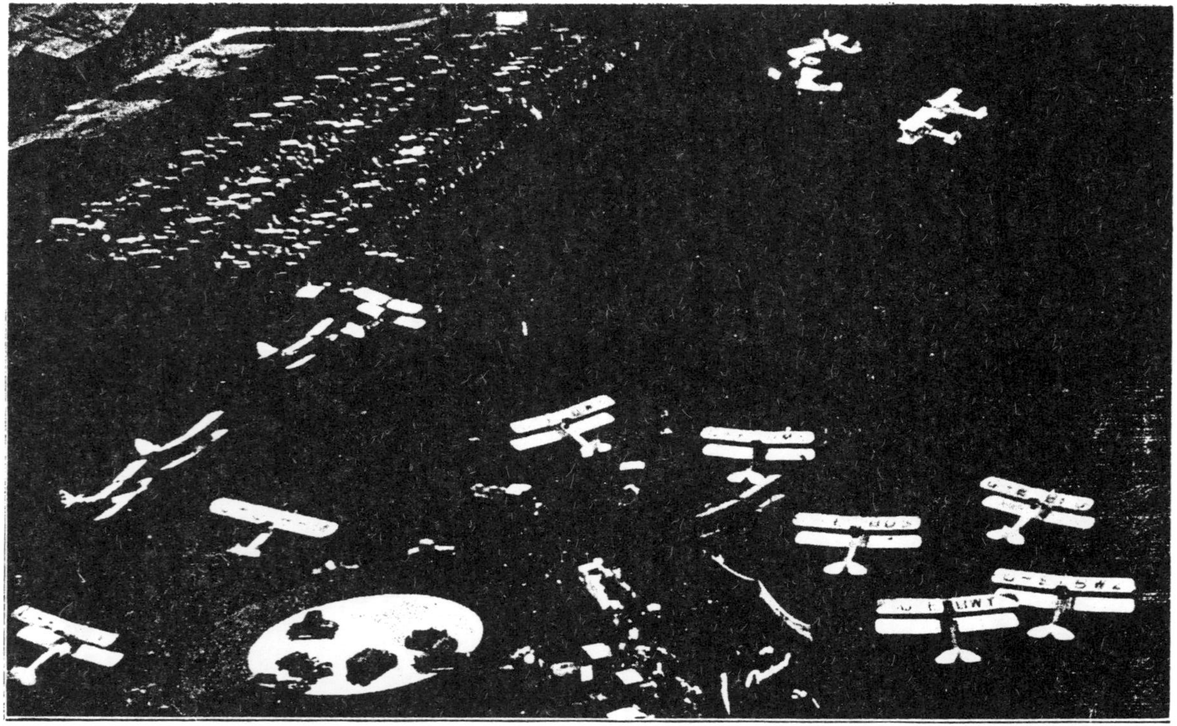

[*Bristol Times and Mirror*

FILTON AERODROME DURING THE WESSEX CLUB'S PAGEANT.

Prince George opens Whitchurch Airport, 31st May 1930

THE GREAT DAY

The airport may not have been ready by January 1st 1930 but all was well for the opening on the 31st May, and what a magnificent day was planned. A royal to perform the ceremony, a pageant organised by the Bristol and Wessex Aeroplane Club whose members were past masters at arranging such events, a great spectacle for the public, and a chance to have a joy-ride (how the meaning of that word has changed), in an aeroplane to view the aerodrome from the air. It rained! The writer remembers the day well and cannot resist describing his own experience of it. As a boy of twelve he set off from Bedminster with friends to walk up over the common on Novers Hill to cross the fields towards Whitchurch. We had never been as far as the site of the Airport but knew it was just a few miles. It did not just rain, it poured. Alas, we abandoned our journey and returned home soaked through.

Prince George, HRH the Duke of Kent, arrived by air at Bristol, and was driven to Whitchurch to open the Airport. He was accompanied by the Lord Mayor of Bristol and the French Ambassador, it was the beginning of the Bristol - French week. The opening ceremony took place from the roof of the clubhouse where the Prince was introduced to many notables. His speech was broadcast nationally and amplified for the spectators in the various public enclosures.[19] There followed a display by many commercial aircraft. Every effort was made to attract as many people as possible. There were three enclosures, one on the north of the aerodrome, with an admission of a shilling (5p), another at the south of the Airport at half a crown (12.5 p), and the "Royal" enclosure where the charge was ten shillings (50p).[20]

Joy riding was always a feature of air displays in those days. Not only were they available on the opening day (Saturday) but also on the Sunday following. Imperial Airways provided flights in an Air Liner, but only from the Royal and half a crown enclosures; Cornwall Aviation Co. Ltd. provided flights from the shilling enclosure in a three seater Avro 504, another famous World War One fighter. It was also stated in the programme that, "Should there be sufficient demand one Avro will also work from the Royal [and] 2/6 enclosure".[21] The Cornish company had been considered for a much bigger role at the new Bristol Airport but it lacked capital. During the negotiations no doubt attention was also given to the possibility of conflict with the Aeroplane Club.[22] The Cornwall Aviation Co. Ltd. returned to Bristol for a "Flying Circus" which performed every day during the period 3rd to 14th June 1931, again flights were offered "from 5/-". Whitchurch was shunned and the circus took place at Hanham, "at [the] tram terminus".[23]

From the opening souvenir, using both text and advertisements, it is possible to build a full picture of the Airport at that time. A map shows the only access was the narrow Bishopsworth - Whitchurch lane, but the line of the projected inner and outer ring roads are shown; both edged boundaries of the Airport. These were only completed after the Second World War, so other roads were built later to improve the situation.[24]

The great theme of many of the articles described a picture of Bristol in the tradition of its sea, railway and road connections now extending to the air. The possibilities of freight and passenger transport were not forgotten but the greater vision was of private aeroplanes, even to the point of rivalling the motor-car. One contributor in describing, "Buying your own aeroplane", states, "In practice it is easier and safer than running one's own motor car." [25] You could buy a single seater aeroplane for £400, a two-seater for £600, and a three-seater with enclosed cabin from £850. You could expect 20 miles per gallon on a small plane, find the oil consumption "almost negligible" and garage it in one of the lock-up hangars at Whitchurch for eighteen shillings(90p) per week.[26] The enthusiasts of flying of those days genuinely envisaged private flying developing in a short time to numbers far, far higher than the present time, some sixty-five years later. There is no doubt that the advent of the Second World War prevented the growth of private flying, and later the high cost precluded its use for ordinary people. An aeroplane for sale could be seen in the showroom and one advertiser offered to accept a motor car in part exchange.

"Learning to fly is a coming thing, you cannot afford to be left out", was the appeal to members of the public to join the Bristol and Wessex Aeroplane Club as a pilot-member. Entrance fee was three guineas (£3.15p) and the annual subscription the same. For this all the facilities of the Airport and the Clubhouse were available to the member. Social members only paid two guineas annual subscription with no entrance fee. It was suggested that the club offered, "the cheapest means of learning to fly", which included, "the use of Club aeroplane, Gypsy Moth, the services of a first-class instructor, petrol and oil provided, and you are insured against liability. The inclusive cost of this is £2 per flying hour." The prospective pilot was also promised, "Fourteen hours of flying instruction should see you a qualified pilot, with an Air Ministry "A" licence, for an outlay of £28 spread over say two months".[27]

A picture of the front cover of a well-used copy of the official souvenir booklet.

The back cover of the booklet.

Perhaps a sobering thought would be that a labourer at that time could earn as little as two pounds per week before stoppages.[28] In spite of that comment it cannot be denied that the Bristol Club, and others like it, provided a reserve of pilots for the country.

The weather marred much of the fun and games of the momentous opening of Bristol's Airport and the programme had to be curtailed, but the whole country, and beyond, was made aware of the entry of the West Country into the whole new field of air transport. What was then available at Whitchurch may appear inadequate to modern eyes but was in fact as good as what could be found in any other British provincial city. The joint management of the City and the Bristol and Wessex Aeroplane Club, under Capt. L. P. Winters, was ideal for the time. See Appendix .

THE EARLY '30s

The Airport Management Committee as a policy adopted a principle of "limited expenditure" during its early years,[29] yet, there was a steady, if slow, growth in facilities. The boundaries of the City and County were extended to include the area of the Airport and in the first year the road at the entrance was improved giving access to two new houses for resident staff. In the same year an agreement was reached with Airwork Ltd. to provide facilities for the repair, inspection and maintenance of aircraft. Its very large, original hangar can still be seen, which, with extensions built during the Second World War, form a part of the present Whitchurch Sports Centre.

The first part of the inner ring road was built and appropriately named, "Airport Road", a title that is still in use although it may well confuse a stranger. The public house, at the junction of the new road and the Wells Road was called, "Happy Landings". From the Airport Road to the terminal buildings of the Airport a narrower, concrete, private road was constructed along the eastern boundary of the aerodrome. The link with Bristol was now complete with a well-marked and speedy access to Bristol's joint railway station at Temple Meads. An enclosure for the public to view the flying was provided.[30]

Improvements were carried out to the Clubhouse to provide better social and catering facilities, the latter to attract flying visitors, and what was an essential requirement, a pilots' room.

This was a place with tables and maps for the pilots, both local and visiting, to plan their routes.[31]

Flying was on the increase throughout the country and Bristol became an important part of both the public and private sectors of air transport. It became an important stage in King's Cup Race every year from 1930. The following table shows passenger and flight figures from 1932 to 1935, they do not include activities of the local Club and other flying schools using the airport.[32]

	1932/3	1933/4	1934/5
Aircraft from other Aerodromes	792	694	3,546
Passengers	1,636	2,604	5,500

Most of the passengers would have been the result of the growing number of airlines using the Airport, one in particular had great local significance.

Above, left) Capt. L. P. Winters, Manager of the Airport and Secretary of the Club. (Right) Capt. P. Bartlett, Instructor of Flying. (Below) The Club's machines (*Bristol Times and Mirror*).

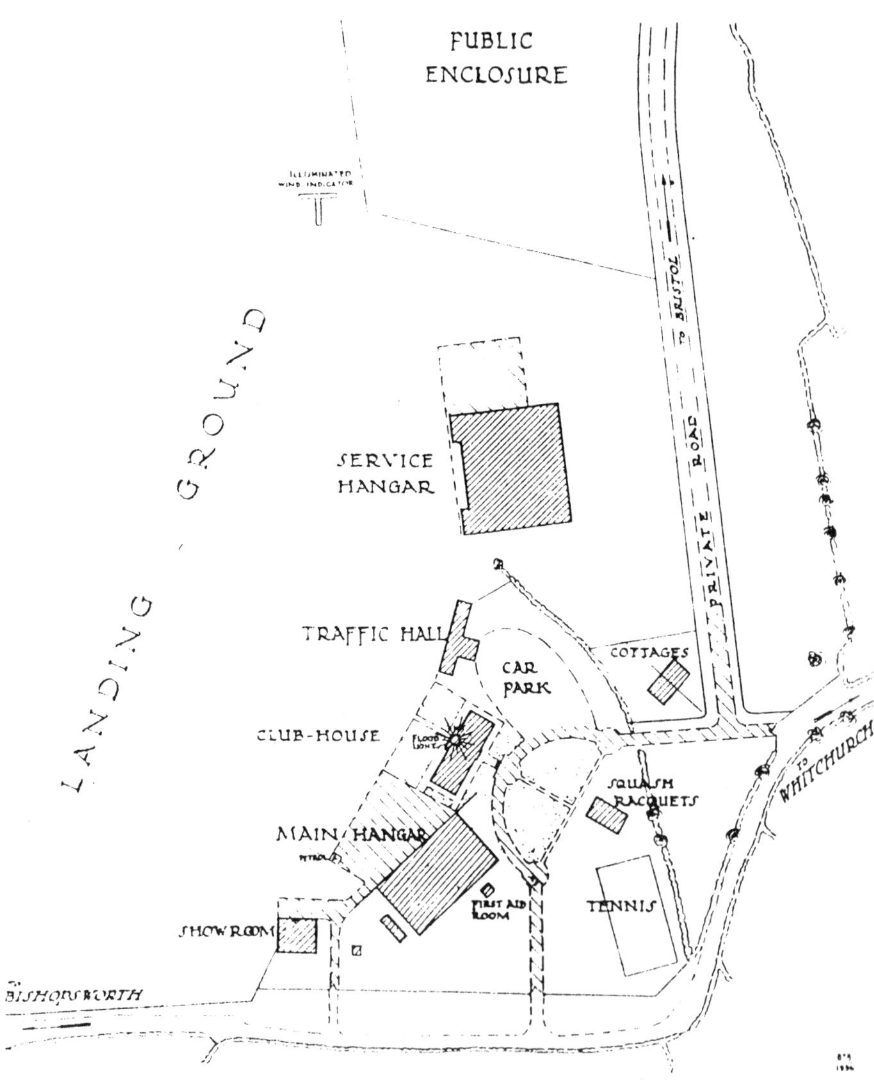

Plan of the "nerve centre" of Whitchurch Airport in the 1930's. The flood light was not in use until 1936. See the picture on the following page.

The main buildings of the Airport in the early 30's. Note the absence of the flood light and look particularly at the shape of the Airwork hangar in the top, left-hand corner. A copy-right photograph from the M. J. Tozer collection, used with permission.

Whitchurch Sports Centre.
The original hangar can clearly be seen forming a major part of the buildings.

THE WEST'S OWN AIRLINE

One of the first provincial airlines was Norman Edgar(Western Airways) Ltd. It was not officially formed until 7th September 1933, it had taken delivery of a DH Dragon, G-ACJT, a week earlier. Norman Edgar, he was involved in aviation as a boy as already stated, took the lease of the showroom at Whitchurch from Merlyn Motors in September 1931 and with some members of the Bristol and Wessex Aeroplane Club started a charter service from the Airport called, "Bristol Air Taxis". On 26th September 1932 he started a Whitchurch (Bristol) to Splott (Cardiff) twice-daily air service. The plane used was the DH80 Fox Moth G-ABYO which carried three passengers. The fare was eight shillings (40p) single and twelve shillings (60p) return.[33] There can be nothing but admiration for those early pioneers of air passenger transport, who took such financial risks when capital and running costs were high and the number of passengers that could be carried so low. Early local leaders of Western Airways were Lord Apsley (Chairman), Norman Edgar (Managing Director), and other directors, Lady Apsley, Mr. H. Crook (of the Kleen-e-ze Brush Company), Mr. K. Maconachie and Mr. L. Ivan Arnott. Major E. Cadbury, (MD J. S. Fry and Sons) and Capt. Douglas Wills, (of the Imperial Tobacco Co.) were among the principal shareholders.

With the purchase of more planes the Airline was able to extend the route to Bournemouth and still later, May 1935, a

Bristol - Le Touquet and Paris route was started. This line was later taken over by Crilly Airways. DH Dragon and Rapide aircraft, with their six to eight seats, were the planes most used for passenger carrying by the smaller airlines.[34]

Western Airways was much involved with the provision of an aerodrome at Hutton Moor Weston-super-Mare, Norman Edgar was the consultant to the Council. As early as 1933 he flew a party of councillors from Weston sands to view the area. However Weston Airport was not built and licensed, at first for Dragons only and, "flown by experienced pilots who are familiar with the characteristics of the aerodrome", until 1936. Those "experienced pilots" in fact built their own control tower, using for the "glasshouse" top, "half of an old seafront cab shelter from Weston-super-Mare". The new Airport was only seven minutes from the sea-front at Weston and the Splott Airport at Cardiff was so near the city that the Weston - Cardiff run was soon very popular, in fact 1,700 passengers used the Weston Airport before the official opening on 25th June 1936.[35] The air-ferry from Weston to Cardiff continued with ever increasing passenger numbers until 1939 when, during the Whitsun period, 2,555 passengers were carried from Weston Airport. This was a world record beating Imperial Airways 1,200 passengers on the London - Paris service during the same period.[36]

Western Airways progressed and expanded with their headquarters at Weston-super-Mare. In 1937 the Straight Corporation Ltd. invested in Western Airways and the following year they took the Company over. Keeping the headquarters at Weston they developed subsidiary bases at Bristol (Whitchurch), Exeter, and Plymouth Airports.[37] During the Second World War Western Airways, like all other airlines, came under the control of the Air Ministry, thus making a major contribution to the defence of the country. After the War the Weston - Cardiff service again became popular but stopped as soon as the Splott Aerodrome was closed and the new Cardiff Airport was constructed on its present site.

During the War, and after, aircraft engineering and manufacture of parts developed at the Weston-super-Mare Airport and Western Airways became a part of the holding company, "Airways Union". Western Airways continued for several years operating from Weston as a charter company and carrying out repairs and maintenance of air fleets as well as being associated with the aircraft works. Flying ceased at Weston about 1980.

The Bristol Airport Bus with a Western Airways Dragon.
Photograph © M. J. Tozer Collection

BRISTOL AIRPORT RECEIVES NATIONAL RECOGNITION

On 14th May 1935 an important report was read to the Bristol City Council from the Airport Committee. It was pointed out that if the policy of "limited expenditure", that had been carried out over the previous five years were to continue, "not only would the Airport not progress but it would not maintain the important place it [had] already attained in the scheme of aviation in this country". The aircraft and passenger figures for 1932 to 1935, given above, were included in the report and the Airport Committee asked for the Council's consent in the provision of wireless and meteorological stations, and lighting equipment for night landings.

Details were given as follows:

> *"Your Committee have been able to induce the Air Ministry to agree, subject to satisfactory financial arrangements being accepted to set up at the Bristol Airport a Wireless Transmitting, Receiving and Direction-finding (Adcock) station of a mobile character which will form part of the national network of civil aviation facilities which is being established at a limited number of centres throughout the country. The Air Ministry state that the question of making the station a permanent one is left for decision at*

some later date and also that it is proposed to make Bristol a meteorological forecasting centre in connection with the same national network after the Wireless Station has been established. Regarding financial arrangements, the Air Ministry point out that the department is not authorised to accept the principle that the expenditure involved should fall on public funds. They are, however, in a position to undertake the services in question during the present state of development of civil aviation provided that a contribution is received from the Council based upon one-tenth of the estimated cost of upkeep to the Department; this arrangement to remain in force if air traffic continues to justify it, until 1939, when the question will be renewed."

The cost of the Council's contribution was estimated at £275 p.a. but only £165 p.a. would have to be paid until such time as the Meteorological Station would be ready. The Council also had to pay for buildings and provide the necessary extension of the electricity supply at an estimate of £900. The importance of the provision of lighting equipment for night landing was stressed to the Council, particularly as Air Mail services had already started at the Airport. The proposal was for the provision of a flood light and an illuminated wind indicator "at a cost not exceeding £900." The Council approved of the expenditure.[38] This decision of the Government to select Bristol for the improved facilities, in spite of "the strong competing claim of another local authority",[39] showed that Whitchurch now had full national recognition.

The Night Landing Flood-light mounted on the Club roof.
9 K.W. 1,200,000 candle power.

1936

!936 was a most important year for Bristol Airport, which now had the latest radio equipment, many airlines using the aerodrome, another route with the opening of Weston-super-Mare Airport and the start of the Bristol - Dublin service. This was the year when the Bristol Corporation took over the full control of the management of the Airport. Capt. P. L. Winters was re-appointed Airport Manager but he was now an officer of the Corporation.[40] Other Corporation staff included a Hangar Foreman, various groundsmen and two junior clerks, one permanent and the other temporary for the summer only. Their main function was the assistance of the Manager in the control of air traffic, consequently they worked early and late shifts from morning until just after sunset.[41]

Some concept of the advance of passenger air transport by 1936 can be obtained from the map and details following, taken from Bristol Airport publicity material. It can be noted that the term used extensively today, "scheduled services", was already in use. Various companies, Western Airways, Railway Air Services, Crilly Airways and Aer Lingus Teoranta [Irish Sea Airways] all used Bristol. Direct services were available to Birmingham, Bournemouth, Brighton, Isle of White, Cardiff, Croydon [by the end of 1936], Dublin, Southampton, Liverpool and Manchester. Arrivals and departures were so arranged that connections made it possible to travel from Bristol by air to and

REGULAR AIR SERVICES

TIME TABLE AND FARES ON APPLICATION.

Companies operating Regular Services from Bristol :—

IRISH SEA AIRWAYS — Telephone Bristol 41165

(Operated jointly by WEST COAST AIR SERVICES LTD.
and AER LINGUS TEÓRANTA)

RAILWAY AIR SERVICES LTD. — ,, Bristol 41165

CHANNEL AIR FERRIES LTD. — ,, Bristol 41165

The above-mentioned Companies operate scheduled services between Bristol and :—

Direct Services to—

Birmingham, Bournemouth, Brighton, Bembridge (Isle of Wight), Cardiff, Cheltenham, Croydon, Dublin, Exeter, Gloucester, Liverpool, Manchester, Plymouth, Ryde (Isle of Wight), Southampton.

Connecting Services to—

Belfast, Blackpool, Bradford, Channel Islands, Glasgow, Isle of Man, Leeds, and Portsmouth.

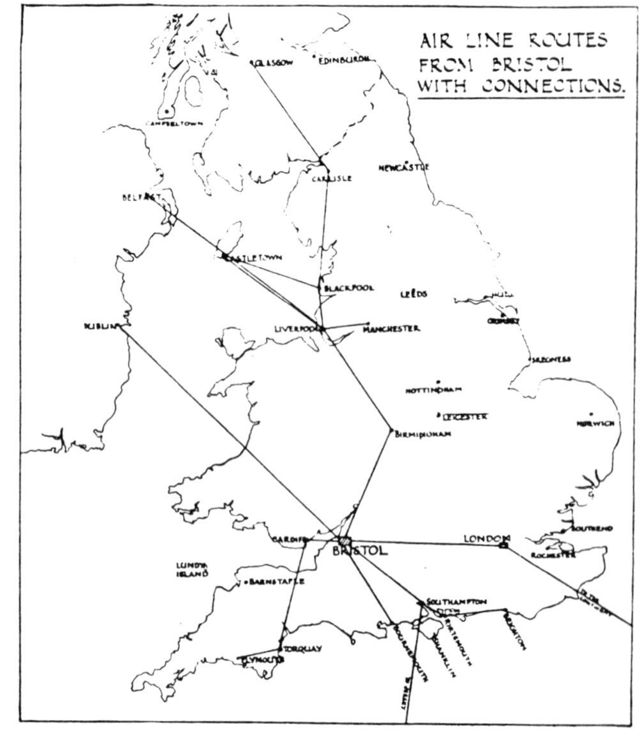

from Belfast, Blackpool, Bradford, Channel Islands, Glasgow, Isle of Man, Leeds and Portsmouth.

All this may well give the impression, particularly when thinking of a modern airport, that there would have been an enormous amount of work for such a small staff. That was not the case as flying in those days was very much controlled by weather, in any case many regular routes were only run during the summer. There were odd times, of course, when the airport was very busy, but often there were other times when there was very little to do. Perhaps some concept of Bristol Airport in those days can be gathered from the daily duties of the "traffic assistants".

The first duty of the junior clerk on his arrival just before eight o'clock was to run up the Civil Aeronautical Flag, and then look around at the weather. The Meteorological Station was still to come, so weather reports had to be made by any officer of the Airport. It was only a report that had to be prepared, not a forecast. The only "instrument" available was a wind direction indicator, observations, therefore, had to be estimated. A cynic may suggest they were guessed. Visibility was judged by checking with known distances, the furthest point that could be seen was the hill beyond Long Ashton, if that was very clear then visibility was reported as being over six miles. Wind speed was obtained by comparisons with the Beaufort Scale, or how the wind sock was filling, the direction came from the wind-vane. Knowledge of the height and amount of cloud at various levels was very important for incoming pilots, that also had to be estimated. It was also usual to include a brief description of the weather as it was at the time and during the previous two hours. Each day, someone who had urgent need of knowledge of the weather at Whitchurch was the pilot at Dublin, who would be flying the Dublin to Bristol service. The information would be converted to a code, known as "Q code", and sent by the radio-telegraph, recently installed at both Bristol and Dublin, to Baldonnel (Dublin) Airport.[42] A longer weather report was then prepared and converted by code into three groups of five figures, this was then sent by telegraph, via the phone and with priority, to the Air Ministry. Later a list of weather reports, including Whitchurch, was broadcast on 1200 metres. Lists of weather stations, with columns to fill in details, were provided. It was the traffic assistant's duty to see that these were filled in and made available on the pilots' notice board, together with other up-to-date weather information.

The radio station, with the Air Ministry Civil Signals operators, was housed at the north side of the Airport; the connection to the control room was by means of a services over-land telephone. All messages had to be written down on special forms both in the control room and the radio station. Radio operators were not permitted by Air Ministry regulations to initiate, or act on, any of the messages. They were mostly ex-R.A.F. servicemen, needless to say the eighteen year old traffic assistants were glad of their advice from time to time. The homing device, which made navigation possible when flying in cloud or bad visibility, was entirely the responsibility of the operators.

The telephone switchboard for the Airport was in the control room and was manned by the assistant on duty. The room was in the front of the Traffic Hall, although not raised, its windows afforded a good view of the whole of the flying area. This enabled a record to be kept of all arrivals and departures,[43] a landing fee of one shilling(5p), had to be collected from all pilots who had not made previous arrangements for payment. All airlines negotiated special terms with the Bristol Corporation. When the weather was fine, and at weekends, perhaps the majority of landings and take-offs were by pupils and members of the Bristol and Wessex Aeroplane Club. B.A. Swallows had been purchased by the Club at the beginning of the year;[44] these planes, with their low, wide, single wing, made a magnificent sight as they approached the airfield and landed, little wonder they had been called, "Swallows". Frequent visitors also were the Tiger Moth Bi-planes from the Flying School of the Bristol Aeroplane Co. Ltd. They were painted in a bright yellow colour. No record was kept of Club and School aeroplanes using the Airport facilities.

The busy times of the day were in the mornings and later afternoons when the planes of the various airlines made their scheduled arrivals and departures. Passengers were able to use the facilities of the Traffic Hall if they were required to change planes. In those days all the passengers had to be weighed as well as the luggage and freight. Most airlines had an office, with a representative, in the Traffic Hall.

Incidents occurring are remembered against the routine of the day, not that any day was dull in the field of ever growing air transport. On one occasion a message was received from an aircraft that its engine cowling was loose, there were fears of a large sheet of metal drifting earthwards. Information was radioed back to the plane that, "a ground engineer was standing by". The plane landed and all was well. The radio operators reported that the incident was almost treated as a "Mayday", all other stations remained off air to give Bristol Whitchurch and the aeroplane exclusive use of the radio band. On another occasion during 1936 Whitchurch was visited by the American all metal Vultee V-1A that was making a record breaking attempt of the flight between New York and Croydon. After meeting a thunder-storm over the Irish Sea the pilot lost his bearings and because of a shortage of fuel he was forced to land in a field in Carmarthenshire where it remained overnight. Its arrival at Whitchurch was of no avail as there were no stocks of high-octane fuel. There were many phone-calls across the Atlantic and eventually it was found that the petrol was available at Filton. B.B.C radio bulletins had kept the public aware of what was happening so when the plane did arrive at Filton it was greeted by "crowds of sight-seers".[45]

As in previous years, again in 1936, Whitchurch was chosen to be a staging post for the Kings Cup Race. The whole Airport took on a gala atmosphere the management of which, and the hospitality, was the responsibility of the Bristol and Wessex Aeroplane Club. The members of the Club also had their usual Garden Party during that year.

The commonest airline passenger-carrying planes at that pre-war time that would have been seen at Whitchurch were the D.H. Dragons and Rapides; the Rapide was an advanced version of the Dragon, it had tapering wings and was capable of a longer range. Bus services, the Bristol one started in 1934, were available from Bristol and Weston-super-Mare to the two Airports. The ordinary citizen could now travel from the City to the seaside, or the holiday maker at Weston-super-Mare visit the tourist attractions of Bristol, and combine their days out with an air trip in a modern, if small, air-liner. Flying for the first time, in a comfortable seat with a good view of the ground, was a thrilling experience. Rising from Whitchurch, no doubt with some apprehension, viewing Dundry church tower from above, and having a birds-eye view of the water tanks at Barrow, would have been new experiences indeed. The grand view of the Bristol Channel, and the yet more apprehension on landing, would have completed the thrills of the day.[46]

The reader will understand how thoroughly the author enjoyed those exciting days when he was eighteen years of age and the temporary "traffic assistant". Instructions were very much service style and very clear. If in doubt the traffic assistant will contact the Airport Manager before taking action, in the absence of the Airport Manager he will consult the Secretary of the Club or the Chief Flying Instructor. If unable to obtain advice he will use his own discretion. On one memorable afternoon that in fact did happen.

The tower of Dundry Church on the ridge can just be seen on this small photograph. Being 800 ft high it was sometimes used to judge the height of very low clouds. The trees and buildings on the edge of what was the air field were not present in the 1930's.

It was a terrible day with poor visibility, low cloud and very little flying; the weather was not expected to improve. The writer was very much on his own. The weather suddenly improved, although the clouds were still low, and the afternoon flying started up again. As the planes from Birmingham, Cardiff and Southampton were scheduled to arrive at Bristol at the same time there was, although slight, a potential of collision. Through the radio station instructions to fly at varying heights were given, all was well. The author's hat did not fit for days!

On another occasion it was dusk and a plane was heard overhead. This traffic assistant had always wanted to see the massive 1,200,000 candle-power working, he threw the huge switch making the landing area ablaze with light. Alas the pilot on landing refused to pay the one pound night landing fee as he had not ordered a night landing. The usual shilling landing fee was accepted with some grace and a receipt given.

The limit of visibility from the old control room at Whitchurch, the western end of the long Ashton ridge can be seen so visibility would have been reported as being "over 6 miles". The vast open space in the foreground leaves little doubt that this was once an airfield.

The staff "cottages" remain but now in private occupation.

THE IRISH CONNECTION

The Dublin to Bristol airline, from its beginning, was looked upon with great affection, not only by most of the Airport staff, but also by many leading citizens of Bristol. Trade between Bristol and Dublin dates from early medieval times and the Irish authorities had been seeking an air link with Britain for many years. The possibility of Ireland becoming a staging post for an American - European airway had had early consideration. Aer Lingus had been planned as a state airline, which it became, but the start was in co-operation with Olley Airways, particularly for the Liverpool service.

On 19th May 1936 the new Civil Signals Radio station was opened at Dublin (Baldonnel) Airport, on 22nd May Aer Lingus Teoranta (Irish Sea Airways) was incorporated, and just five days later on 27th May 1936 the first flight from Dublin to Bristol was made.[47] The plane was the D.H. Dragon, EI ABI, it was blessed by the Irish Air Corps chaplain, the Revd. William O'Riordan.[48] The pilot was Capt. Oliver Eric Armstrong, he was an Olley pilot then but was soon appointed as Chief Pilot of Aer Lingus. The passengers were W.H.Morton, a director, Mr. and Mrs. T. Fitzherbert of Dublin, T.J.O'Driscoll and Mrs. Sean O'uHadhaigh. At Bristol the pilot and passengers were greeted by the Lord Mayor and the Chairman of the Airport Committee, Alderman A. A. Sennington. Unfortunately that first plane was late, this was due to problems with the radio, understandable when in fact it was only delivered to Dublin from the factory on the previous evening.[49]

The Dublin to Bristol flight was always intended to double as the first part of a journey to London. Had the Dragon been re-fuelled at Bristol and then continued to Croydon the passengers would have had a long journey to the centre of London. In fact a large limousine met the plane at Whitchurch and drove the passengers, who were continuing to London, to Temple Meads Station where they caught the "Bristolian" to Paddington. At that time it was the fastest regular passenger train in the world. This arrangement continued until the Autumn when the Dragon was changed for the superior DH 86a Rapide, EI ABK and the journey was extended by air to Croydon.

The daily routine for the Irish flight for that first summer started, as already described, with the Bristol weather report radioed to Dublin. Later Bristol would receive a radio message from Dublin starting as, "EI ABI ARMSTRONG DEP DU 0900 HRS etc.........". This would have been followed by another message, this time from the plane, to the effect that it was, "Passing the Brecon Beacons and changing over". From that point the pilot would have kept in touch with the Bristol Radio Station, the operator, using the Adcock system, would find the bearing of the plane from Bristol, and calculated a "reverse bearing" for the pilot to navigate. All being well the last message received from the pilot would have been, "Passing Avonmouth and winding in". The pilot would then wind in his trailing aerial and land on sighting the Airport.

With an airline established the routine of each day would have been very much the same, however there were exceptions. On one occasion an "overcast" weather report was given to Dublin, the plane left at the usual time and it came out of the clouds to land at Whitchurch on time. One of the passengers was a sea captain, his experience of going into clouds at Dublin and seeing no land until the plane reached Bristol Airport amazed him. He no doubt thanked Capt. Armstrong and then added words, with suitable nautical language, to the effect that he had sailed the world but how the pilot had brought the "ship" to Bristol he would never know. Another incident caused disappointment for all concerned. The morning flight was made in record time as there was a tail wind. An actress was booked on the evening flight to Dublin, she was appearing that evening in the Dublin Opera House. The wind did not change direction and remained high. Capt. Armstrong worked out every conceivable route where he could land and re-fuel but it was unavailing as he would have run out of fuel. Had the flight been successful Aer Lingus would have benefited from the publicity. The eventual use of the Rapide no doubt made such a situation less likely to be repeated.

On another occasion the report was received that Capt. Armstrong had sighted Avonmouth, then the plane must have entered cloud as it was not seen for some time. Readers can imagine the relief when the roar of the engines was heard overhead and EI ABI made its usual perfect landing.

Capt. Armstrong made a report to Aer Lingus on the number of passengers carried in the early days of Aer Lingus. It was:

26th May to 6th July 1936

Dublin to Isle of Man	***171 passengers***
Isle of Man to Liverpool	***41 passengers***
Dublin to Bristol	***47 passengers*** [50]

The services continued until the outbreak of war when flights ceased, "outside of Ireland". The Bristol flight was resumed very early in the War with a considerable increase of passenger numbers.

The passengers on the very first flight from Dublin are greeted at Whitchurch by Alderman Sennington, Chairman Airport Committee. Photo Aer Lingus.

The Aer Lingus restored Dragon flying at Badminton. The ex-Western Airways plane was re-registered EI ABI and named "Iolar".

The "crash wagon" may not look sophisticated to modern eyes but it was well equipped for its day. Photograph from the M. J. Tozer collection.

FURTHER IMPROVEMENTS

During the period 1937 to 1938 the Bristol Council, with financial help from the Government, approved major improvements at the Airport. Boundary lights were installed, and the drainage of the flying area was improved. The major cost was the acquisition of large areas of land around the Airport to prevent obstruction by the private construction of high buildings.[51] At the time it was hoped that the cost would be offset by using the land for recreational purposes. In April 1939 approval was given for a new control room. The wider area was also intended for making use of the developing electronic aids for landing and taking off in poor visibility. After 1939 there are no accounts of the Airport Committee in the Council minutes.

A restored BA Swallow. The very successful training machine used by the Bristol and Wessex Aeroplane Club in the late 1930's.

WAR

This study was intended to finish with the outbreak of the Second World War, it has told the story of Bristol's part in keeping apace with provincial and overseas air transport. However, as far as civil transport was concerned the taking over of the airlines and the Bristol Airport by the Government was in fact the zenith of the early history of the aerodrome at Whitchurch. Plans were made well before the outbreak of hostilities, Bristol was to be the centre for civil aviation and the headquarters of Imperial Airways and British Airways. All was secret, much has been revealed since. Writing in 1989 about Bristol and describing a photograph, Hugh J. Yea states, " 'Here be Dragons.' Fortunately, in view of what was to happen later, the foresight of the Bristol City Fathers made possible the opening of the municipal airport at Whitchurch, only 3 miles from the city centre in 1930",......"the airport played an important role in air communication within the United Kingdom, being particularly well placed in relation to journeys involving water crossing - either short-haul across the channel to Cardiff, or on the longer sea crossing to the Channel Islands via Bournemouth, or from 1936 between Bristol and Dublin". The heading, "Here be Dragons" was an apt description of Bristol Whitchurch Airport at the beginning of the war. It has been reported that DH Dragons and Rapides came from all over the country to Bristol, causing accommodation problems particularly in times of high winds.[52]

Since hostilities ended much has been written about Whitchurch during the War years, particularly by members of the crews on the Imperial Airways flights, Bristol to Lisbon.

A former flight steward, Ronald Meredith, writes, *"These flights, operating in wartime, flew thousands of miles unescorted, with the loss of only one aircraft due to enemy action. All on board perished including actor Leslie Howard, who played Ashley Wilkes in Gone With the Wind."* and, *"the most hairy part of the flight was missing Dundry Tower on foggy November days on return."* J. R. Williams, a catering apprentice with B.O.A.C., describes how busy Whitchurch was during the years 1941 - 42, and states, *"The British Overseas Airways operated from there, operating flights to Ireland to link up and bring passengers from the US. The aircraft was a De Havilland Frobisher, made from wood and fabric. I saw and spoke to many a film star - Edward. G. Robinson, Bob Hope, Francis Langford etc"*.[53]

A NOSTALGIC EPILOGUE

Whitchurch, returned to the City as a civil aerodrome after the War, it ceased to be an airport in 1957 when it "moved" to Lulsgate. It is now Hengrove Park, a vast open space with its wartime runway still in position. The surroundings have changed, but the two staff cottages are still intact, and the original Airwork hangar, with wartime extensions, can also be seen within the structure of the Whitchurch Sports Centre.

At the time of the preparation for the publication of this book an announcement was made in the Bristol Evening Post regarding the new leisure centre at Whitchurch. Work should be well underway when the book is first available. The new complex, developed by T.H.I. of Chester and financed by Sun Life Assurance, is an example of Bristol City Council joining with private enterprise. It is also thought to be the first time a pension fund has been involved in a leisure investment.

The new centre will consist of a 48 room Travel Inn, a quarter million pound adventure playground for children, a 14 screen cinema, a 900 seat bingo hall , and two restaurants. The buildings will surround a parking area for 900 cars. The 21 acre site is on the north side of the old airport and access will be from the outer ring road (A4174). Opening is planned for Christmas 1997. The profit that Bristol Council will make from the scheme will be used for the development of the remaining 225 acre Hengrove Park. Paul Smith, the chairman of the city land and properties sub-committee, has stated that he sees the scheme as creating a new confidence in South Bristol. A second phase with the possibility of a ten pin bowling alley and a virtual reality centre, is already in the planning stage.

Filton, as a result of the well known post-war Brabazon project, now has the longest runway in the British Isles. An application by British Aerospace for the use of the aerodrome for scheduled flights, mainly for business use, has been turned down, the result of an appeal is awaited. There has been opposition from local residents and the Lulsgate authorities; opinions are very much divided. At the time of writing feelings are running high as there is even a movement to ban the return of the much loved Concorde to Filton for popular, pleasure flights.

At Weston-super-Mare there is similar discussion over the future use of the abandoned airport and industrial complex. The transport plan for the former Avon area suggests that Weston Airfield should be reopened for business/recreational use.[54] At the present time a massive Sunday market is held there. The airfield is remembered by name as motorists drive around the new, nearby, "Airport Roundabout".

A much happier story comes from Aer Lingus, the Dragon EI-ABI, has been re-created. The original, sold to Channel Air Ferries, was shot down by enemy action on a flight from Lands End to the Scillies. With real Irish sentiment, in 1945, Aer Lingus purchased a Dragon formerly belonging to Western Airways and restored it. It was also re-registered as EI-ABI. The plane has visited the Bristol area and there was a plan to re-enact the first flight of 27th May 1936 from Dublin to Bristol on the 50th anniversary, but it would have been Bristol, Lulsgate.[55]

Alas the weather was too bad for the plane to attempt the journey.

Aer Lingus not only still functions as a state airline but it has retained its original name. At the present time it is possible to fly Aer Lingus to Dublin from Bristol (Lulsgate), the passengers can also fly on to New York or Boston having passed customs and immigration examinations before leaving Dublin.

A final thought, before it is too late and the remaining space is built upon, it would be a lovely sight to see EI-ABI landing at Whitchurch for the last time. Nostalgia indeed!

Exercising dogs on the war-time runway at Whitchurch, a popular present day pastime.

Inside the Dragon "Iolar" on its local flight. Roger Bennett of Radio Bristol and other flying enthusiasts enjoy the unusual experience.

The restored Aer Lingus "Iolar" flying near Yate on its way from Badminton via Filton to Lulsgate. Edwin Shackleton, who arranged the flight, is seated behind the pilot.

APPENDIX A
1930
The Airport Committee

The Lord Mayor (Councillor Walter Bryant)

Alderman A. A. Senington (Chairman)

Alderman A. Dowling

Alderman H. C. Woodcock

Councillor R. Ashley Hall

Councillor F. Berriman

Councillor H. R. Griffiths

Councillor J. E. Jones

Councillor R. F. Lyne

Councillor A. L. H. Smith

Councillor J. Owen

The Management Committee

Representing the Corporation

Alderman A. A. Senington

Alderman A. Dowling

Councillor A. L. H. Smith

Representing Bristol and Wessex Aeroplane Club

Councillor R. Ashley Hall (Chairman Management Committee)

A. Taylor, Esq.

C. F. Uwins, Esq.

L. M. Leaver, Esq.

Airport Manager

Capt. L. P. Winters

REFERENCES, NOTES AND BIBLIOGRAPHY

References and Notes

1	Anon	*A History of the Bristol Aeroplane Company Limited.* 1960. Typescript Filton Public Library p. 1.
2	Ibid	p.2.
3	Pudney J	*Bristol Fashion - Some account of the earlier days ofBristol Aviation.* Putman. London 1960. pp 49 & 53.
4	Ibid	pp 67 & 69. Fairlawn Avenue was demolished a year or so before the writing of this book.
5	Regan R J	*A brief history of No. 50 (County of Gloucester) Squadron.* c.1964. Typescript Filton Public Library.
6	Pudney J	*Bristol Fashion.* Putman. London. 1960. pp 95 - 99.
7	Barnes C.H.	*Bristol and Wessex Aeroplane Club - The first fifty years.* Typescript Archives B & W Aeroplane Club Bristol Airport (Lulsgate). Undated probably c 1977. Pages not numbered.
8	A Member	*Light Aeroplane Clubs No.7 The Bristol and Wessex Aeroplane Club.* Series of articles in the magazine "Air".June 1928. It is unfortunate that the name of the author of this excellent article, describing the early days of the Bristol and Wessex Aeroplane Club, is not known.
9	Ibid	p 22.
10	Ibid	p 23.
11	Ibid	p 24.
12	Barnes C.H.	*Bristol and Wessex Aeroplane Club - The first fifty years.*Typescript Archives B & W Aeroplane Club Bristol Airport (Lulsgate) Undated probably c 1977. Passim.

13 Wilson Harold *A Prime Minister on Prime Ministers.* Michael Joseph. London 1977. Picture p. 204.

14 Personal recollections.

15 Bristol Council Minutes of Council Meetings. Printed and bound. Bristol Central Library. Volume Nov. 1927 - Oct. 28.

16 Ibid Vol. Nov. 28 - Oct. 29. p 205.

17 Ibid p. 335.

18 Bristol Council Manuscript minutes of Airport Committee 29th December 1929. Bristol Record Office.

19. Bristol Council *Bristol Airport England Official Opening Day Souvenir.* Glovers Advertising. Bristol 1930 pp 34 - 35.

20 Ibid p 30.

21 Ibid p 31.

22 Chapman Ted *Cornwall Aviation Company* Glasney Press Falmouth 1979 p 16

23 Ibid p 51.

24 Bristol Council *Bristol Airport Opening Day Souvenir.* p 11.

25 Ibid p 54.

26 Ibid p 55.

27 Ibid p 56.

28 Personal recollections. The writer was a pay clerk in the Aircraft Industry 1936 - 8.

29 Bristol Council Minutes of Council meeting, 14th May 1935. Vol. Nov. 34 to Oct. 35. p 256.

30 Ibid Vol. 1932 to 1933. p 37.

31 Ibid Vol. 1933 to 1934. p 181.

32 Ibid Vol. 1934 to 1935. p256.

33 Simons G. M. *Western Airways - The West Country Airline.* Redcliff Press. Bristol. 1988. p 5

34 Ibid p 6.

35 Ibid p 7.

36 Ibid p 16.

37 Ibid p 8.

38 Bristol Council Minutes of Council Meetings. Vol. 1934 - 35. p 256.

39 Ibid p 257.

40 Bristol Council Minutes. Vol. 1935 - 36. p 189.

41 Personal recollections. The writer was the temporary junior clerk during the summer of 1936. Where references are given they confirm and, in some cases, revise his memory. He was offered the post the following year but in those uncertain days of unemployment it was refused.

42 Share B. *The Flight of the Iolar - The Aer Lingus Experience 1936 - 1986.* Gill and Macmillian Dublin. 1986. p 30. Confirmation of the daily receipt of a weather report in "Q" code by radio-telegraph from Bristol Whitchurch.

43 Personal recollection. In spite of extensive inquiries the large register of arrivals and departures has not been found, either in the Bristol Record Office or the present Bristol Airport (Lulsgate). Perhaps, not surprising, in view of the upheaval of war and the movement to Lulsgate.

44 Barnes C. H. *Bristol and Wessex Aeroplane Club - The first fifty years.*

45 Pitchers T. *Bristol Airport.* Article c.1978. The writer can well remember this incident, he spent some time putting through phone calls to America for the pilot and passengers - the first time he had ever made a call abroad. The article provided information he did not recall, the type of aircraft, the radio coverage and the crowds at Filton.

46 Personal recollection The writer's first flight, and last for at least fifty years.

47 Share B. *The Flight of the Iolar - The Aer Lingus Experience 1936 - 1986.* Gill and Macmillian Dublin. 1986. pp 5/6.

48 Ibid p 26

49 Ibid Passim

50 Ibid Passim

51 Council minutes1936,1937,1938 and 1939.

52	Yea Hugh J.	*September 1939: Westwards to Whitchurch*. The Putnam Aeronautical Review. IssueThree. 1989.
53	Letters	Bristol Evening Post 5th November 1994.
54	Avon County	*Transport Plan for the Avon Area 1994 - 2013* Avon County Council Publicity June 1995 Air Transport Section.
55	Crossland N.	*Irish Eyes Smile for a Happy Return.* Bristol Evening Post May 1986.

Bibliography

Harris W.L.	*Filton, Gloucestershire Some Account of the Village and Parish* W.L. and N.L. Harris Bristol 1981.
Simons Graham M.	*Western Airways the West Country Airline* Redcliff Press Bristol 1988.
Pudney John	*Bristol Fashion* Sub title *Some account of the earlier days of Bristol Aviation* Putnam London 1960.
Share Bernard	*The Flight of the Iolar - The Aer Lingus Experience 1936* - 1986 Gill and Macmillan Dublin 1986.
Chapman Ted	*Cornwall Aviation Company* Glasney Press Falmouth 1979.
Bristol Council	*Bristol Airport England Official Opening Day Souvenir* Three sections 1 Annual Handbook 2 Official Programme 3 Bristol and Wessex Aeroplane Club. Glovers Advertising Bristol 1930 Bristol Record Office.
Avon County	*Transport Plan for the Avon Area 1994 - 2013* Avon County Council Publicity June 1995.

Primary Sources

Bristol Council	Minutes of Council Meetings 1927 - 1939 Printed and bound Bristol Central Library.
Bristol Council	Minutes of Airport Committee Handwritten in minute books. Bristol Record Office.

Unpublished Papers etc.

Gillett Steph	*The Bristol Aircraft Industry* Module III Diploma Course in Industrial Heritage Copy Filton Public Library.
Anon	*A History of the Bristol Aeroplane Company Limited 1960* Typescript Filton Public Library
Regan R.J.	*A brief History of No 501 (County of Gloucester) Squadron* c. 1964 Typescript Filton Public Library
Anon	*60 years of Bristol Aircraft* c. 1970 Typescript Filton Public Library
Wakefield Ken	*Bristol (Whitchurch) Airport 1930 - 1957* Typescript prepared for Whitchurch Local History Society Undated
Barnes C.H.	*Bristol and Wessex Aeroplane Club - The first fifty years.* Typescript Archives B & W Aeroplane Club Bristol Airport (Lulsgate) Undated probably c 1977 Pages not numbered

Magazines and Newspapers

A Member	*Light Aeroplane Clubs No.7 The Bristol and Wessex Aeroplane Club* Series of articles in the magazine "Air" June 1928
Yea Hugh J.	*September 1939 - Westwards to Whitchurch* Putnam Aeronautical Review Issue Three October 1989
Letters	Bristol Evening Post 5th Nov. 1994

Crossland N. *Irish Eyes Smile for a Happy* Return Bristol Evening Post May 1986

Bernadette Kitterick *Hengrove Park Developments* [Bristol] Evening Post 7th Nov. 1996

A "Moth" at a Wroughton "Fly In". This was a popular training plane for schools during the 1930's.

POSTSCRIPT

For someone who knew the traffic hall as it was at Whitchurch in 1936 and then, some fifty years later, to see the arrivals area at Bristol Lulsgate for the first time, surprise would be a very mild word to describe the sensation. And then to be conducted around the whole of the airport buildings, including visits to the control and radar rooms and a meeting with many of the staff, wonderment would be a better word to summarise his feelings. That was the experience of the author when the visit was arranged in 1986. For the fifty years celebration of the airline Aer Lingus he had given some help in describing the early days at Whitchurch.

All who use Lulsgate know of the changes that have taken place and the expansion of the airport since1986. It is also known that more are to come, including the diversion of the A38. However other changes are being considered, the possibility of the use of Filton as an airport for scheduled and business flights, and even, the suggestion of building an entirely new airport on a greenfield site near the Second Severn Crossing.

These last few words are an appeal for negotiations for these and similar matters to be speeded up. Would Bristol have ever had an airport if the civic fathers in the late 1920's had taken so long to make decisions? The records show that they were very expedient, thus giving Bristol a good start in passenger air transport. Most people would want Bristol to maintain its position as leaders in aircraft design and have regional air transport facilities, both for passengers and freight, as good as any in the country.